WEIRD SEA CREATURES™
THE OCTOPUS

Miriam J. Gross

The Rosen Publishing Group's
PowerKids Press™
New York

For Gillian

Published in 2006 by The Rosen Publishing Group, Inc.
29 East 21st Street, New York, NY 10010

First Edition

Editor: Daryl Heller
Book Design: Albert B. Hanner

Photo Credits: Cover © Stuart Westmorland/Getty Images; p. 5 © Jeff Rotman/naturepl.com; p. 6 © Reinhard Dirscherl/SeaPics.com; p. 9 © Marilyn & Maris Kazmers/SeaPics.com; p. 10 © Mark Conlin/SeaPics.com; p. 13 © Mike Severns/SeaPics.com; pp. 14, 21 © Fred Bavendam/Peter Arnold, Inc.; pp. 17, 18 © John C. Lewis/SeaPics.com.

Library of Congress Cataloging-in-Publication Data

Gross, Miriam J.
The octopus / Miriam J. Gross.
     p. cm. — (Weird sea creatures)
Includes bibliographical references and index.
ISBN 1-4042-3188-9 (library binding)
1. Octopuses—Juvenile literature. I. Title. II. Series.

QL430.3.O2G76 2006
594'.56—dc22

                              2004022437

Manufactured in the United States of America

CONTENTS

INTRODUCING THE OCTOPUS

The octopus is a soft, rubbery sea creature with no bones. This animal can change color, squirt, or spray, ink, and speed away from trouble. The octopus is able to do these things because this creature is so **vulnerable**. With no hard shell to **protect** it, the octopus **evolved** with unusual features and **intelligence** that keep it safe from dangers that hide in the sea.

Scientists have classified, or labeled, the octopus as a cephalopod, which means "head-foot," in Greek. This describes the way the octopus' eight arms attach directly to its head. The cephalopod family also includes squids and cuttlefish.

Cephalopods are part of a larger group of animals called mollusks. Mollusks are soft-bodied invertebrates, or animals without backbones. Clams, oysters, and snails are all mollusks, and most of them have hard shells to keep themselves from harm. The octopus, however, has evolved so that it does not need a shell.

This octopus is swimming in the Red Sea in Egypt. The small cup-shaped circles underneath the octopus' tentacles, or arms, are called suckers. Adults of some large species of octopuses can have more than

tentacles

head

funnel tube

suckers

An octopus has good eyesight. This helps it find food, even at the bottom of the ocean where there is little light. Unlike people the octopus does not see colors, such as blue or green. The only colors an octopus sees are black and white.

OCTOPUS BASICS

The octopus' body is made up of a large head and eight armlike **tentacles**. The total size of an octopus' head and tentacles can be as small as 2 inches (5 cm) or as big as 25 feet (8 m). The bottoms of the tentacles are covered in suckers. These suckers are made up of tiny **muscles** that tighten to grab hold of surfaces. The suckers also have fine senses of touch, taste, and smell. These allow the octopus to find food and avoid enemies in the dark.

The octopus' head includes a covering of skin called the mantle, which surrounds the brain, stomach, and three hearts. The mantle forms a space, called the mantle cavity, around the octopus' gills. Gills are the **organs** that most sea creatures use to draw oxygen from the water to breathe. The octopus breathes by pulling seawater into the mantle cavity with a tube, called a funnel tube, which is located at the base of its head. Water can also be pushed out of the funnel tube to move the octopus through the water. The octopus' eyes and a mouth with a sharp **beak** are also at the base of its head.

WHERE THE OCTOPUS LIVES

Octopuses like to keep their homes clean, so they push the remains of their meals outside their dens. Divers can find octopus dens by looking for these piles of shells.

There are more than 250 **species** of octopuses. From the cold Arctic to the warm **tropics**, octopuses are found in nearly every ocean of the world. Some octopuses live on the sandy ocean floor, as far down as 23,000 feet (7,010 m), or almost 4.5 miles (7 km) below the surface. Most octopuses, however, live in less deep waters, such as **tide pools** and underwater caves closer to shore.

Octopuses are **solitary** creatures. They make their homes, called dens, in small spaces like cracks, or small openings, in rocks or the gaps in a **coral reef**. An octopus can squeeze, or fit, its soft body into tiny spaces that are about the size of its eye. They can live in garbage, such as cans, pots, and bottles. Octopuses often build walls of rocks in front of the entrances to their dens. These walls have only a small opening. This helps the octopus hide from **predators** that are too large to enter the dens.

This octopus has made a den inside a conch shell. A conch, like the octopus, is a mollusk. Once a

The octopus is hard to find in this picture. By changing its texture and color, this octopus has camouflaged

THE OCTOPUS GOES INTO HIDING

The octopus can change its appearance up to 20 times per minute. Most octopuses are colored with spots of grey, white, and brown. An octopus can change the color of its skin using a group of color cells called a chromatophore system. Different cells become larger or smaller to create different colors. Tiny muscles control the size of these cells. When the muscles pull the color cells wider, the cells get lighter. When the muscles relax, or let go, the color cells get darker.

Sometimes an octopus will lose a tentacle in a fight with a predator. If this happens the tentacle will twitch, or move, on the ground for several minutes. The twitching tentacle distracts the predator while the octopus makes its escape. The octopus will soon grow a new tentacle.

The octopus changes color to **camouflage** itself to hide from predators or to sneak up on **prey**. Changes in color might also show the way an octopus is feeling. It often turns white when frightened and red when it is angry. The octopus can also change the texture, or feel, of its skin from smooth to bumpy. In this way the creature can look like a rock or a piece of coral.

WHAT'S FOR DINNER?

Octopuses are carnivores, which means they eat meat. They like crabs, clams, shrimp, fish, fish eggs, and squid. They will sometimes eat other octopuses as well. One way the octopus hunts is by sitting still while blending in with, or matching, nearby rocks or seaweed. An unlucky fish or crab that does not see the octopus may swim right into the octopus' arms. The octopus then grabs the prey with its suckers and **paralyzes** it with **venom** from its beak. An octopus can also find food by poking its long tentacles into the cracks between rocks, where small prey might be hiding.

The meat in animals such as snails can be hard to get at because of their hard shells. In this case the octopus first grabs the snail with its tentacles. Then it drills a hole in the shell with its radula, which is a strong, bumpy tongue inside the octopus' beak. The octopus next injects venom through the hole. This causes the snail's soft body to come apart from its shell, so the octopus can eat the animal.

This octopus is attacking a cone snail. Using its radula an octopus takes several hours to drill a hole through a shell such as this one.

After letting out a cloud of ink from its funnel tube, this giant Pacific octopus uses jet propulsion to speed away. Some species of octopuses can shoot an ink cloud that is shaped like an octopus. This tricks the predator into thinking that the cloud of ink is the octopus.

How an Octopus Escapes

The octopus' soft body is an appealing snack for many animals. Sharks, moray eels, sea otters, seals, and bottom-dwelling fish, such as lingcod and halibut, all eat octopuses.

If a predator is chasing an octopus, the octopus can squirt out a cloud of ink from its funnel tube. The ink cloud forms a smoke screen. The predator is blinded by the ink while the octopus escapes.

When it needs to swim away quickly, the octopus fills its mantle with water and then blasts the water out through its funnel tube. The stream of water gives the octopus a boost and pushes it along backward. This blastoff is called jet propulsion. The octopus can also hide by fitting itself into a tight space. With no bones the octopus can fit into any opening that is bigger than its beak and brain case. This means it can fit into a space that is about the width of its eye.

People have eaten octopuses for thousands of years. One common way to catch octopuses dates from the ancient Greeks and Romans. Fishermen lower jugs into the water and rest them on the seafloor. Once the octopuses have adopted the jugs as dens, the fishermen pull them out.

MATING HABITS

A mother will not leave the cave to hunt during the time her eggs are growing. Therefore, she may starve to death before the babies are born. If this happens, another animal will probably eat the young.

Octopuses spend most of their lives alone. They come together only to **mate** in the last half of their lives. When octopuses are ready to mate, the male will wait outside a female's den. He may have to fight off **rival** males. The male mates many times in a season. The female, however, is only able to have babies once in her life.

After she mates, the mother finds a safe cave or den for her eggs. She may lay up to 50,000 eggs at a time. The mother will often glue the eggs together and stick them to the top of a cave or to a rock to keep them from being carried away by the water currents. The mother cleans the eggs and supplies them with extra oxygen by squirting water on them with her funnel tube. She never leaves her eggs. The mother may also build a wall of stones at the entrance to her cave.

This female southern blue-ringed octopus guards her eggs. The egg cases are transparent, or clear. The tiny eyes and tentacles of the octopuses can be seen through each case.

The baby velvet octopus on the right has almost completely hatched, or broken free, from its egg case. The octopus on the left has not yet hatched. The babies will not stay together. Once they hatch water currents will carry them off to different locations.

LIFE CYCLE OF AN OCTOPUS

In places where the weather is cold, it can take six and a half months for the eggs of some octopus species to hatch. However, in areas where the weather is hot it can take fewer than six weeks. Baby octopuses look just like tiny adults. They can already change color, squirt ink, and use jet propulsion.

Upon hatching, the babies are carried away by the sea. For the first few months of their lives, they drift as **plankton** on ocean currents. This leaves them vulnerable to hungry fish and whales. If they live past this stage, they grow quickly and get 20 percent bigger every day. The young octopuses then settle on the ocean floor or in caves near the shore.

Many species of octopus will reach their full adult size after one year. Most octopuses live only a year or two. Some species, such as the giant Pacific octopus, may live up to five years.

THE LARGEST AND THE DEADLIEST

The giant Pacific is the largest of the nearly 250 species of octopuses in the world. They may grow to be more than 20 feet (6 m) long and can weigh more than 100 pounds (45 kg). The giant Pacific is found most often near the coast of Seattle, Washington, where huge forests of **kelp** grow from the ocean floor. They eat fish or anything else they can catch. Some people in Washington even claim to have seen an octopus reach out of the water to grab a seagull off a rock.

At 4 inches (10 cm) long, the blue-ringed octopus of the Pacific waters near Australia and Japan is one of the smaller octopus species. However, this species of octopus is also the deadliest. The bite of a blue-ringed octopus has a poison called tetrodoxin that is strong enough to kill a person in a matter of hours. A blue-ringed octopus would only bite a person if it felt it was in danger. Therefore, it is best to leave these brightly colored creatures alone.

This giant Pacific octopus lives in the Pacific Ocean near British Columbia, Canada. When this octopus

OCTOPUSES AND PEOPLE

Octopuses often escape from aquarium tanks. Octopuses can figure out how to lift the lids and move the handles of their tanks. They have also been known to sneak into the tanks of other animals, eat the other animals, and then sneak back into their own tanks before their keepers can catch them.

For many centuries octopuses were considered monsters, creatures that waited in the ocean and used their tentacles to attack ships and pull sailors to their deaths. In recent years scientists have been able to study octopuses, both by diving into their natural surroundings and by studying them in aquariums. They have discovered that octopuses are gentle, shy creatures that will only attack people to **defend** themselves.

Scientists have also learned that octopuses are by far the smartest of all invertebrates. In tests octopuses have run mazes and solved puzzles. In one case an octopus was given a glass jar with a lobster inside. After several tries the octopus learned that it needed to remove the jar's lid to get at the lobster. This talent for learning has helped the octopus find food and avoid danger in its ocean home. By learning about octopuses, people can help them stay safe.

GLOSSARY

beak (BEEK) A pointed piece of the octopus' mouth that is hard and sharp.

camouflage (KA-muh-flaj) To hide by using a color and a pattern that matches one's surroundings.

coral reef (KOR-ul REEF) Underwater hill of coral.

defend (dih-FEND) To guard from harm.

evolved (ih-VOLVD) Changed over many years.

intelligence (in-TE-lih-jintz) The ability to learn from the past and use this knowledge to make choices for the future.

kelp (KELP) A large, brown seaweed.

mate (MAYT) To join together to make babies.

muscles (MUH-sulz) Parts of the body under the skin that can be tightened or loosened to make the body move.

organs (OR-genz) Parts inside the body that do a job.

paralyzes (PAR-uh-lyz-iz) Causes a loss of feeling or movement in a part of the body.

plankton (PLANK-ten) Plants and animals that drift with water currents.

predators (PREH-duh-terz) Animals that kill other animals for food.

prey (PRAY) An animal that is hunted by another animal for food.

protect (pruh-TEKT) To keep from harm.

rival (RY-vul) Having to do with those who try to get or do the same thing as another.

solitary (SAH-leh-ter-ee) Spending most time alone.

species (SPEE-sheez) A single kind of living thing. All people are one species.

tentacles (TEN-tih-kulz) Long, thin growths usually on the head or near the mouths of animals used to touch, hold, or move.

tide pools (TYD POOLZ) Areas of shallow water at the seashore that are surrounded by rock.

tropics (TRAH-piks) The warm parts of Earth that are near the equator.

venom (VEH-num) A poison passed by one animal into another through a bite or a sting.

vulnerable (VUL-neh-reh-bul) Open to attack or harm.

INDEX

WEB SITES

Due to the changing nature of Internet links, PowerKids Press has developed an online list of Web sites related to the subject of this book. This site is updated regularly. Please use this link to access the list:

www.powerkidslinks.com/wsc/octopus/